100 WAYS TO USE WHEAT

All Sugar-Free Recipes

(38 Bread Recipes)

Nature's Greatest Gift

For Your Health

Laura M. Hawkes

ISBN NO. 0-89036-116-9

Typesetting by

Hawkes Publishing, Inc.
3775 South 500 West
Salt Lake City, Utah 84115
Tel. (801) 262-5555

CONTENTS

I

Introduction

The Purpose of Food

The food we eat is to build, repair and operate our bodies. If our food intake is inadequate we cannot be at our best. Poor nutrition can cause us to tire easily, be nervous, cross and ill. On the other hand, a person who has a well balanced diet is healthy, happy and full of energy.

We need proteins, carbohydrates, fats, mineral salts, vitamins and water. Proteins are body-building foods for growth and repair. They are found in meat, eggs, cheese, milk, grains and vegetables.

Carbohydrates are energy foods. They include starches and sugars found in fruits, vegetables and grain.

Fats are also energy foods that produce heat as well. They are found in butter, meat, fish and vegetables.

Mineral salts regulate the body processes and are found in vegetables, grains, nuts, milk, eggs and meat.

Vitamins are body regulators and promote reactions, without which full nutrition is impossible.

Water is the solvent of all foods and is part of every living cell. It is called a food because it is a regulating substance in nutrition.

If we eat a **balanced** diet of the foods available we will be contributing to our good health. It has been said that we, as a nation, eat too much salt, too much sugar, too much fat and too much meat. I am sure that this is so. We need all the things mentioned above. We need carbohydrates but when we eat candy and desserts we are getting too much sugar. Fresh fruits will supply the carbohydrates needed in our diet. French fried potatoes, potato chips, fried meat and other fried foods, and rich pastry give our bodies more fat than they need. Eating large amounts of meat is hard for our bodies to handle. To eat plenty of fruits and vegetables, a little meat and other protein, will give us a well balanced diet and what we need to have good health and energy. There is no need to take vitamins unless our doctor advises it.

Fruits and vegetables are at their best when they are fresh. Manufactured foods, convenience foods, foods with preservatives and additives should be avoided.

A food that is of great worth to us is whole wheat—nothing added and nothing taken away.

THE VITAMIN B COMPLEX

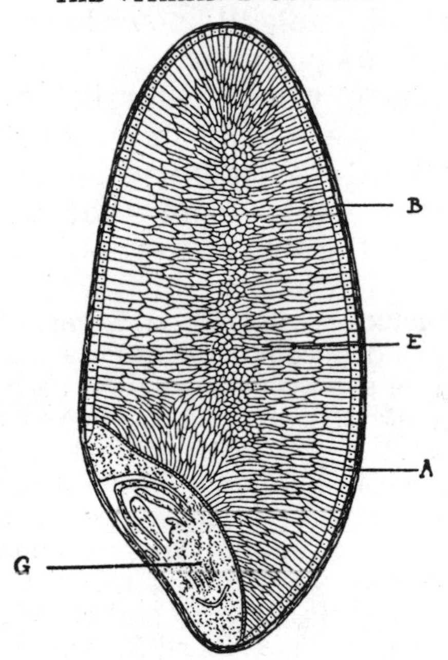

FIG. 27—STRUCTURE OF A GRAIN OF WHEAT.

Diagram of a longitudinal section through a grain of wheat, showing: *B*, pericarp, forming the branny envelope; *A*, aleurone layer of cells forming the outermost layer of the endosperm removed with the pericarp during milling; *E*, parenchymatous cells of the endosperm; *G*, embryo or germ. (*Vitamins, Medical Research Council*, p. 159.)

The whole kernel, bran and germ included, is very necessary for good health. If you have good quality wheat and grind it yourself just before making bread, you will know that you have kept all the nutrients possible.

The value of wheat in our diet is discussed by John A. Widtsoe and Leah D. Widtsoe, in their book, **The Word of Wisdom**, which I recommend. I will insert a topic from that book, entitled "Structure of the Wheat Kernel."

> **Structure of the Wheat Kernel**. There are three well-defined parts of the wheat kernel: (1) the bran or outer covering, (2) the germ, in which the new life is located, (3) the internal portion or endosperm, which forms the bulk of the kernel. The bran consists of four layers, the three inner coats of which are rich in protein (aleurone), minerals and vitamin B. The germ, under proper conditions of warmth and moisture will germinate, hence is the living part of the grain. It is very rich in vitamins B and E, in protein, fat and mineral matter. The endosperm or central part is composed of starch with about one-sixth as much gluten (protein) enmeshed in cellulose (woody fiber), but is devoid of vitamins and contains very little mineral substance. The organic phosphates, essential for the nutrition of brain and nerves, and the inorganic minerals, calcium and other phosphates concerned in the formation and growth of bone and teeth, are found mainly in the bran and germ, while the endosperm, from which white flour is made, is very poor in these

important salts, containing less than one-tenth of the amount found in bran. For all these reasons "wheat for man" is good advice. But to secure the full value of wheat, man should eat the whole grain.

The average difference in composition among whole wheat, white flour (the endosperm) and the bran (the outer covering), is shown in the following table:

	Wheat Kernel	White Wheat Flour	Wheat Bran
Water (moisture)	12.00	13.50	13.00
Mineral Matter (ash)	2.00	0.40	5.80
Protein (nitrogenous matter)	12.00	11.00	15.40
Fat (ether extract)	2.00	1.25	3.60
Starch and Sugar (carbohydrates)	70.20	73.60	53.20
Cellulose (crude fiber)	1.80	0.25	9.00
	100.00	100.00	100.00

Explanations:

C. cup
tb. tablespoon
t. teaspoon

Honey is sweeter than sugar. If a recipe calls for 1 C. sugar and you wish to use honey, use about 3/4 C.

II
Wheat Storage

Wheat is the most important item to store. Hard dark winter (turkey red) wheat is recommended as the best wheat to store. Store in a cool place if possible. If it is stored in a basement we should make sure that the wheat container doesn't touch the cement floor or the walls. If cement touches a metal container for a period of time it will rust. Use tin cans with tight-fitting lids or plastic containers especially made for food storage (with tight-fitting lids). To insure against weevil, add 2 tb., no more, of pure ethylenedichloride to a 5 gal. container of wheat. Pour on the wheat and cover with a tight-fitting lid. When the container is opened the ethylenedichloride will evaporate. This does not change the quality of the wheat or its uses in any way. In planning for a year's supply of food for your family, you should have about 300 to 365 lbs. of wheat per family member. An adult would need about 1 lb. per day, depending on what other food you have.

III
How to Sprout Wheat

Select good untreated wheat seeds, discard any broken or bad seeds and wash thoroughly. Cover with water, twice as much water as seeds, and soak for about 10 hours, in warm weather and from 12 to 16 hours in the cold weather. Drain off the water and put in a sprouter, or a pan or can. Cover with warm water and drain several times a day. Keep in a warm room in a dark place or keep the container covered for about three days. The sprout should be about the length of the wheat seed. If you want the sprouts to be green, expose them to the light.

IV
Gluten

Gluten is the protein that is found in wheat. It is very nutritious and much less expensive than other proteins. It is tasteless and can be flavored as you wish. It is very easily made.

Gluten Recipe

Mix 12 C. Whole Wheat flour with 6 C. cold water and knead for about 10 min. Then wash and drain. Take about a cup of the mixture and hold it under the cold water tap, letting a small stream of water touch the bread as you work it in your hands. Have a pan underneath to catch the water. Work the dough until the water runs clear. Let the dough drain in a colander. Repeat this process until all the dough has been taken care of. You should have about 3 or 4 C. of gluten.

Grease a cookie sheet, place gluten on it to about ½ in. thickness. Bake for 15 min. at 350 degrees. Prick any raised areas with a fork and bake 15 min more. When the gluten has cooled you may grind it with a grinder or cut it in strips or squares.

Raw gluten may be stored in the refrigerator for about 2 days and cooked gluten may be stored about 5 days. It may be frozen raw or cooked. Gluten is not a complete protein and needs to be used with another protein such as meat, cheese, milk, eggs, nuts or soy products.

The baked gluten may be ground and used as ground beef, pork, chicken or in desserts according to the flavoring and seasoning that you use with it.

The water that was collected while washing the gluten should be allowed to stand for several hours. It can be used in drinks, gravy, soup, sauces and puddings. If you have no other use for the starch water it can be used to water plants or animals. It needs to be kept in the refrigerator, as it spoils quite readily. The bran can be used for cereals, pancakes, muffins, cookies or powdered and used in the future. The thick white starch can be used as corn starch.

V
Making a Sour Dough Starter

The starter must be set the night before you wish to use it. If it is replenished every week with flour and water it can last for months, even years.

2 C. flour
2 C. warm water
1 pkg. dry yeast

Combine ingredients and mix well. Place in a warm place or closed cupboard overnight. In the morning put ½ C. starter in a scalded pint jar with a tight lid and store in the refrigerator or cool place for future use. This is a sourdough starter. The remaining batter can be used immediately for pancakes, waffles, muffins, bread or cake.

To use the starter again, place it in a medium-sized bowl, add 2 C. milk, 2 C. flour. Beat well and set in a warm place to develop overnight. In the morning the batter will have gained ½ again its bulk and be covered with air bubbles. It will have a pleasant yeast odor. Now, set aside ½ C. sponge in a tight jar in the refrigerator for your sourdough starter for next time.

VI

Bulgar Recipe

1. Wash wheat in water and pour off the water.

2. Add enough water to cover the wheat and boil slowly until the water is absorbed and the wheat is tender, about 40 min.

3. Spread a thin layer of wheat in a pan or on a cookie sheet and place in the oven at 200 degrees until the wheat is dry and cracks easily.

4. Wet the surface of the wheat and rub the kernels between the hands to loosen and remove the chaff.

5. Crack the wheat in moderate sized pieces with a mill or grinder or a hard object. Some of the wheat may be left whole. Store in a covered can or other container.

Bulgar may be used in any recipe calling for cracked wheat or bulgar.

Bulgar for Cooked Cereal

½ C. bulgar, cover with water and add ½ t. salt and cook until tender. This has a very delicious taste and takes but a few minutes to prepare.

VII
Yeast Breads

Whole Wheat Bread

2 tb. dry yeast or 2
compressed yeast cakes
4 level tb. honey
2 level tb. salt
4 tb. shortening or cooking oil
½ C. warm water
2 large eggs
1 C. regular powdered milk
1 C. wheat germ
12 C. Whole Wheat flour
5 C. warm water

Into ½ C. warm water sprinkle 2 tb. dry yeast and allow to stand.

Into a large pan or bowl place the salt, honey, shortening or oil, and 5 C. warm water, add lightly beaten eggs and yeast. Mix dry ingredients together and add gradually, stirring with a large spoon. When the dough is too stiff to stir with a spoon, knead with hands. If it is too sticky add a little additional flour. Cover with a wet cloth and let rise in a warm draft-free spot. Form into loaves and put into four greased bread pans and let rise again until almost double in size. Put in a 400 degree oven and let bake for 10 min. Then lower the heat to 350 degrees and bake until nicely browned on top and bottom (nearly 1 hour).

For 2 loaves of bread, divide the recipe in two.

See page 2 of introduction as pertaining to honey and sugar.

Wholewheat Spoon Bread

1¾ C. warm milk
1 pkg. dry yeast dissolved in 1/3 C. warm water
1 tb. cooking oil
1 to 2 tb. honey
2 t. salt
1 egg beaten
3 to 3½ C. unsifted
Whole Wheat flour

To warm milk add dissolved yeast, egg, oil, honey and salt. Beat vigorously after each addition and let rise at room temperature for 30 min. Stir down and spoon into greased loaf pan. Let rise 30 min. or until not quite double in bulk. Bake 1 hr. at 325 degrees to 350 degrees. Makes one loaf.

Three Hour Whole Wheat Bread

½ C. boiling water
½ C. scalded milk
2 tb. cooking oil
1 tb. molasses or honey
2 compressed or dry yeast cakes
½ C. unbleached flour
3½ C. Whole Wheat flour
1 t. salt

Combine water, milk, oil, honey or molasses and mix. Cool until mixture is warm. Add the yeast cakes and stir until dissolved. Combine the dry ingredients and beat into the first mixture. Transfer to floured bread board and knead until smooth and elastic. Rub all over with oil. Place bowl in pan with enough warm water to heat the bowl. Cover with a warm damp cloth and place in a warm place. When the bread is double in bulk, place on a floured board and knead. Form into a loaf and place in a warm place and let rise until double in bulk. Bake in a moderately hot oven, 350 to 375 degrees, for about 45 min. until brown on top and the loaf can be tipped out without sticking. This makes one loaf.

Whole Wheat Raisin Bread

Follow the recipe for Three Hour Whole Wheat Bread and add 1 C. raisins with the flour before the dough is kneaded the first time. Bake as directed.

Whole Wheat Fig and Nut Bread

Follow the recipe for Three Hour Whole Wheat Bread, and add ½ C. chopped dried figs and ½ C. chopped walnuts with the flour before the dough is kneaded the first time. Bake as directed.

Onion Bread

2 yeast cakes softened
in ½ C. warm water
3 tb. honey
¾ pkg. dry onion soup
2 tb. cooking oil
1½ C. hot water

Add 5½ to 6 C. flour to make a stiff dough. Knead about 5 min., then let rise 1½ hr. Punch down and let rise ½ hr. Divide into loaves and place in pans and let rise until double in size. Bake at 375 degrees for 20 min., reduce heat to 300 degrees, and bake for 25 min. more.

Sprouted Wheat Bread

2 pkg. of dry yeast
½ C. warm water
4 C. warm water
1 tb. salt
2 C. sprouted wheat
4 tb. molasses or honey
4 tb. vegetable oil
1 C. regular powdered milk
6 C. Whole Wheat flour

Sprinkle dry yeast over ½ C. warm water and let stand until dissolved. Add 4 C. warm water, salt, molasses or honey and vegetable oil, then add powdered milk and whole wheat flour. Mix thoroughly then add sprouted wheat and knead, to make a soft elastic dough. Add flour if needed. Cover with a damp cloth and place in a warm place and let stand until it doubles in bulk. Form into 4 loaves and put in greased bread tins. Let rise until nearly double in bulk. Bake at 400 degrees for 15 min. and at 350 degrees for 45 min. or until done.

Unleavened Bread

7 C. warm water
½ t. salt
½ t. honey
½ C. Whole Wheat flour
¾ C. warm water
½ C. flour
6 C flour
1 tb. salt

Put 7 C. warm water in a bread pan or large mixing bowl. Add ½ t. salt and ½ t. honey and ½ C. Whole Wheat flour. Stir, then add enough flour to make a thin batter. Beat well. Place in a warm place and let stand overnight or for about 8 to 9 hrs. Add ¾ C. warm water and ½ C. flour and set in a warm place for about an hour until bubbles appear. Then add 6 C. flour and 1 tb. salt. This should make a good stiff batter. Add more flour if necessary. Sprinkle with flour and set aside to rise. When the batter begins to bubble and break through the flour, make into loaves. Let rise and bake. 400 degrees for 15 min. and 350 degrees for 45 min. You may use Whole Wheat or unbleached flour.

Oatmeal Bread

1½ C. warm water
1 t. salt
1½ C. rolled oats
1/3 C. honey
1 compressed yeast cake
  or dry yeast cake
¼ C. warm water
3 to 3¼ C. unbleached flour
1 tb. cooking oil

Break compressed yeast cake into ¼ C. warm water and let stand or sprinkle dry yeast cake slowly into water and let stand. Mix 1½ C. water, salt and honey. Add oatmeal, then yeast and then unbleached flour. Let rise until double in bulk. Make into a loaf and place in a greased loaf tin or pat into round loaf in round pan and bake 40 to 45 min. at 375 degrees.

French Bread

1½ C. hot water
1 tb. honey
1½ t. salt
1 tb. shortening or oil
1 tb. yeast (1 cake)
4 C. flour (unbleached)
1 egg white with 1 tb. water

Put water, shortening, honey and salt in bowl and let stand until luke warm. Add yeast, then flour and knead until blended.

Put lid or cloth over bowl and let stand for 10 min. Knead again four more times at 10 min. intervals. After the fifth time, roll out to get the bubbles out and roll up like a jelly roll. Brush top with egg white and water mixture and sprinkle with sesame seeds. Cut two or three slashes across top. Bake at 375 degrees for 45 min.

Raisin Bread

For 1 loaf of bread, when you make the recipe for white bread. Add to it 1 C. raisins and knead them into the bread. Dates may be used if you wish. Form into a loaf, let rise and bake at 375 degrees for 30 to 35 min.

Cracked Wheat Bread

For 1 loaf of bread, when you make the recipe for white bread. Add to it 2 tb. honey and 1 C. cracked wheat. Make into a loaf, let rise and bake at 375 degrees for 30 to 35 min.

Potato Bread (2 loaves)

2 C. dry mashed potatoes
1 C. water in which potatoes were cooked
1 tb. salt
½ C. warm water
2 tb. honey
1 tb. shortening or oil
1 cake compressed yeast
2 C. whole wheat flour
3½ to 4 C. unbleached flour

Peel 6 medium-sized potatoes. Cut into pieces and cook in boiling water until tender. Drain the water from the potatoes, but save the water to use as moisture for the dough, and for mixing the yeast. Mash the potatoes; add the potato water, salt, honey and shortening. Sprinkle yeast in ½ C. warm potato water and let stand. When the first mixture has cooled until it is warm, add yeast to this when it is ready. Add flour enough to make a good firm dough. Turn out on floured board and knead until soft and elastic. Cover with a damp cloth and let rise until double in bulk. Form into loaves and place in two greased loaf tins. Let rise again until double in bulk and bake about 1 hr. at 375 degrees.

Oatmeal Potato Bread (2 loaves)

1½ C. potato water
2 C. rolled oats
1 tb. salt
1 tb. honey
2 C. Whole Wheat flour
1 tb. shortening or oil
2 C. dry mashed potatoes
1 tb. dry yeast (1 pkg.)
1 tb. warm water
About 4 C. unbleached flour

Heat liquid to boiling point and pour over rolled oats. Add salt, honey and shortening. Stir and let stand until warm, add potatoes and then proceed as in recipe for potato bread. Let dough rise in pans until it is 2¼ or 2½ times its original bulk. Bake at 375 degrees for about 1 hr.

Casserole Cheese Bread

1 C. milk
1½ tb. honey
½ tb. salt
1 C. warm water
1 tb. cooking oil
2 pkgs. compressed yeast or
2 tb. dry yeast
1 C. grated cheddar cheese
1 C. Whole Wheat flour
3½ C. unbleached flour

Scald milk, stir in honey, salt and 1 tb. cooking oil. Cool to warm. Put 1 C. warm water in large bowl and sprinkle yeast over it. Let stand until yeast is dissolved then add warm mixture, cheese and flour. Let stand until yeast is dissolved then add warm mixture, cheese and flour. Stir until blended.

Cover and let rise in a warm place until more than double about 45 min. Stir batter down and beat vigorously for a short time. Put in 1½ qt. casserole or two 9x5x3 loaf pans. Bake uncovered at 375 degrees for about 1 hr.

Leaven or Maori Bread N.Z.

3 C. unbleached flour
2 tb. or 2 pkgs. dry yeast
3 C. warm water
1½ tb. honey
7 C. unbleached flour
1 t. salt
2 tb. honey

Place water in large bowl and sprinkle yeast over it and let stand until dissolved, then add 1½ tb. honey and 3 C. unbleached flour and let stand for 1 to 1½ hrs. Add the rest of the flour, honey and salt. Mix and let stand for 1 hr. at warm temperature. Put on board and knead for 10 min. Add more flour if necessary. The dough should be quite firm. Divide in 3 loaves or make one large loaf. Put in large covered pan that has been greased and let rise until double in bulk. Bake at 400 degrees for ½ hr. Lower heat to 350 degrees and bake for 1 hr longer, or until done.

Cottage Cheese Bread
(1 loaf)

1 pkg. yeast
¼ C. warm water
1 C. creamed cottage cheese
1 tb. honey
1 tb. instant minced onion
1 tb. butter
2 t. dill seed
1 t. salt
¼ t. soda
1 egg
2¼ to 2½ C. unbleached flour

Place warm water in mixing bowl and add yeast, let stand until soft, combine cottage cheese (which has been warmed) with honey, onion and other ingredients excepting flour. Gradually add flour, beating after each addition. Cover with a damp cloth and put in a warm place and let rise until double in bulk, 50 or 60 min. Stir down with a spoon. Put into a greased round casserole or loaf tin and let rise again until double. Bake at 350 degrees for 40 to 50 min. until golden brown.

Cracked Wheat Bread (5 loaves)

5 C. unsifted cracked wheat flour
6 or 7 C. unbleached flour
2 pkg. dry yeast
3 tb. honey or molasses
1 C. regular powdered milk
4½ C. warm water
2 tb. cooking oil
4 t. salt

Sprinkle yeast over ½ C. water and allow to dissolve (don't stir). Mix dry ingredients and add liquid and other ingredients. Knead until smooth and elastic. The hands may need to be greased often because the dough is sticky. Cover bread with a warm wet cloth and put in a warm place and allow to rise until double in bulk. Knead and allow to rise again. Form into loaves and place in greased loaf tins. Allow to double in bulk. Bake at 400 degrees for 15 min. and then finish baking at 350 degrees.

Swedish Rye Bread

Use the recipe for white bread — delete the honey and add 2 tb. molasses and 2 tb. dark corn syrup and ¾ C. rye flour, 1 C. raisins if desired. Bake at 400 degrees for 30 to 35 min.

Cheese Pimento Bread

2 tb. butter
3 tb. flour
1 tb. honey
2 t. salt
1/3 C. grated cheddar cheese
3 tb. minced pimento
¼ C. warm water
1 compressed yeast
3 to 3½ C. unbleached flour

Put butter, 3 tb. flour, honey and salt in a sauce pan and stir in 1¼ C. milk. Cook and stir over low heat until thickened, then add 1/3 C. grated cheddar cheese and 3 tb. minced pimento. Stir until cheese is melted and cool to lukewarm. Mix together ¼ C. warm water and 1 compressed yeast. Add to the above mixture. Then stir in 3 to 3½ C. unbleached flour. Knead and let rise twice. Shape into a loaf and put in a loaf pan 9x5x3 inches. Let rise to about 1½ times its size for about 30 min. Bake 45 to 50 min.

Whole Wheat Bread (Sour Dough)

2 C. milk
2 tb. butter
2 t. salt
¼ C. honey
½ t. baking soda
¼ C. wheat germ
2 C. Whole Wheat flour
4 C. unbleached flour
1 pkg. active dry yeast or (1 tb.)

Scald 2 C. milk and melt butter and honey in it, allow to cool to luke warm, sprinkle yeast over milk and let stand to dissolve. Add to a quart of sour dough basic batter. Sift into dough 2 C. whole wheat flour and wheat germ. Add salt and soda, stirring gently. Set dough in warm spot, cover with cloth and let set for 30 min. Add remaining flour and stir until bread is too stiff to stir with a spoon. Turn out on floured board and begin to knead with hands. Note flour may vary from quantity indicated. One must gauge by feel whether it is too little or too much. Knead 100 times until dough is light and shiny to touch. Make into loaves and put in greased pans. Pans should be ½ full. Let rise to double in bulk. Set in warm 400 degree oven and bake for 20 min.; reduce heat to 325 and bake until bread shrinks from sides of pans.

White Bread
for 4 loaves

Mix together in large bowl:

4½ C. warm water or milk
4 tb. honey
1½ tb. salt
4 tb. shortening
14 to 14½ C. unbleached flour

Sprinkle 2 tb. dry yeast into ½ C. water and let stand to dissolve. Add yeast and shortening and flour. Add the flour gradually. When you cannot use a spoon, knead with hands. Let rise to double in bulk. Form into loaves and place in four greased pans and let rise again. Bake at 425 degrees for 25 to 30 min. or until done. (For liquid you may use water, milk or potato water. If you use raw milk, scald and cool before using.)

Cornmeal Bread

1½ C. water
1 t. salt
1/3 C. yellow cornmeal
¼ C. honey or molasses
1½ tb. shortening
1 compressed yeast cake
or dry yeast
¼ C. warm water
4 to 4¼ C. unbleached flour

Bring to a boil in a saucepan 1½ C. water, salt and yellow cornmeal. Cool to lukewarm. Crumble compressed yeast cake in ¼ C. warm water or if dry yeast is used, sprinkle slowly into water. Stir molasses or honey and shortening into cornmeal mixture, then add yeast and stir. Add 4 to 4¼ C. unbleached flour. Knead. Let rise until double. Punch down and form into loaf and put in greased loaf tin or pat into rounded shape and bake in round pan at 375 degrees for 40 to 45 min.

Sour Dough Bread

4 C. sifted unbleached flour
(or more)
2 tb. honey
1 t. salt
2 tb. shortening
¼ t. soda
1 or 2 eggs

Set the sponge as usual, saving ½ C. starter. Sift dry ingredients into a bowl, making a well in the center. Add shortening to the sponge and mix well. Add beaten eggs. Pour into the well of flour, add enough flour to make a soft dough for kneading. Knead on a floured board for 10 min. Cover with a damp towel and let rise in a warm place for 2 to 4 hours or until doubled. (If you want the bread to rise faster, add some dry yeast dissolved in ¼ C. warm water to the sponge). Dissolve ¼ t. soda in a tb. of water and add to the dough. Knead it thoroughly. Shape dough into loaves and place in greased bread pans, and set aside to rise. When doubled, bake at 375 degrees for 50 to 60 min.

VIII
Quick Breads

Boston Brown Bread

Mix together:

1 C. rye flour
1 C. Whole Wheat flour
1 t. soda
2 C. sour milk or buttermilk
1 C. cornmeal
¾ C. light molasses

Steam 3 hours. Serve with beans, coleslaw, etc.

Apricot Bread

1 C. dried apricots
2/3 C. honey
2 tb. soft butter
1 egg beaten
¼ C. water
½ C. orange juice
2 C. unbleached flour
2 t. baking powder
½ tb. soda
½ C. chopped nuts

Soak apricots in warm water for 30 min. Dry apricots and cut in pieces with scissors. Cream honey, butter and beaten egg. Sift dry ingredients together and add to egg mixture, alternately with orange juice and water. Add chopped nuts and apricots. Bake in a greased and wax paper lined loaf tin for 55 to 65 min. at 350 degrees. Remove from pan, remove paper and cool on rack.

Apple Loaf

½ C. shortening or cooking oil
½ C. honey
2 C. unbleached flour
1½ tb. baking powder
½ t. salt
1 C. raw, unpeeled apple

Cream butter and honey. Add eggs and beat well. Add dry ingredients that have been sifted twice. Add apples and nuts. Pour into greased loaf tin and bake about 1 hr. at 350 degrees.

Carrot Loaf

2/3 C. honey
1 C. Whole Wheat flour
2 C. unbleached flour
1½ t. cinnamon
1 t. soda
1 t. salt
¾ C. cooking oil
3 eggs
1 small can (abt 6 oz.)
crushed pineapple
2 C. finely grated carrots
3 t. vanilla
1 C. chopped nuts

Place all these ingredients into a large mixing bowl and stir until the ingredients are evenly mixed. Do not stir one of two ingredients at a time.

Grease a loaf pan, place some waxed paper in the bottom and grease that, then put the mixture in the pan and bake 1 hr. and 15 min. at 325 degrees. This makes 1 large loaf or 2 smaller ones. If you make 2 smaller ones, you don't need to bake it as long.

Banana Loaf

2/3 C. honey
2 tb. soft shortening
1 egg
1½ C. milk
½ t. soda
1 C. mashed bananas
1 C. Whole Wheat flour
2 C. unbleached flour
1 t. baking powder
1 t. salt
¾ C. chopped nuts

Mix honey and shortening, stir in egg and mix, sift flour, baking powder and salt together and add alternately with milk and bananas, blend in nuts and place in a well greased 9x5x3 inch loaf pan and bake at 350 degrees, 60 or 7 min. or until wooden pick thrust into center comes out clean.

Cherry Pecan Loaf

2/3 honey
½ C. butter
2 eggs
1 C. unbleached flour
1 C. Whole Wheat flour
1 C. maraschino cherries, chopped
1 t. soda
1 C. buttermilk
1 C. chopped pecans
½ t. salt
1 t. vanilla

Cream butter, honey and eggs until light and fluffy. Sift dry ingredients together and add to creamed mixture alternately with buttermilk. Beat until blended. Add nuts, cherries and vanilla. Place in a greased loaf tin or square pan and bake at 350 degrees for about 1 hr.

Irish Soda Wheat Bread

1 C. unbleached flour
2 C. Whole Wheat flour
1 t. soda
1 tb. baking powder
1 t. salt
2 t. honey
1 tb. butter or margarine
1½ C. buttermilk

In a large bowl combine unbleached and whole wheat flour, soda, baking powder, salt and honey. Work in shortening with finger tips. Add buttermilk all at once and stir with large spoon just until dry ingredients are moistened. Turn out on floured board and knead gently until smooth and rounded up into a ball. Do not knead more than one minute. Place in 9 in. pie tin and flatten to a 9 in. circle about 2½ in. thick. Press a large floured knife across the center of the loaf almost through to the bottom. Repeat at right angles to divide loaf into quarters. (These quarters are called "farls" in Ireland.) Bake at 375 degrees for 35 to 40 min. or until loaf is brown and sounds hollow when tapped.

Orange Loaf

2 C. unbleached flour
1 C. Whole Wheat flour
3 t. baking powder
½ t. salt
2¼ t. orange rind
¾ C. shortening (part butter)
2/3 C. honey
2 eggs
2/3 C. canned milk
1/3 C. orange juice

Cream shortening and honey, add eggs, put dry ingredients in flour sifter and add them alternately with milk. Add orange juice and rind. Put in loaf pan or flat pan if you desire. Bake at 350 degrees until done. Test by thrusting pick into center of cake. When it comes out clean the cake is done. You may add ½ C. nuts if you wish.

Whole Wheat Quick Bread

3 C. sifted Whole Wheat flour
4 t. baking powder
½ t. salt
1½ C. milk
2/3 C. honey
1 egg beaten
1 t. vanilla

Sift dry ingredients together twice. Add milk, honey, egg and vanilla. Pour into greased loaf pan and bake 1 hr. at 325 degrees. (This also makes a delicious nut bread by adding 1 C. chopped nuts.

Gingerbread

1½ C. unbleached flour
1 C. Whole Wheat flour
1½ t. soda
½ t. salt
1 t. ginger
1 t. cinnamon
½ t. cloves
½ C. cooking oil
1/3 C. honey
1 egg
½ C. molasses
1 C. hot water

Mix dry ingredients. Cream oil, honey and molasses. Add beaten egg. Add dry ingredients alternately with hot water. Beat until smooth. Pour into a well greased floured pan 10x14 in. Bake at 350 degrees for 40 to 45 min.

Corn Bread

1 C. yellow corn meal
½ C. unbleached flour
½ C. wheat germ
2 tb. honey
1 C. milk
1 tb. baking powder
1 t. salt
1/3 C. cooking oil
1 egg
¼ C. sprouted wheat

Combine dry ingredients in a mixing bowl, beat egg and milk together and add to dry ingredients along with oil. Add sprouted wheat and pour into a greased 8 in. square pan or in muffin tins. Bake at 400 degrees for about 25 min.

Zucchini Bread

1 large zucchini grated (2 C.)
¾ C. cooking oil
1 t. salt
1 t. soda
½ t. baking powder
2 t. cinnamon
3 t. vanilla
2/3 C. honey
3 eggs
2 C. unbleached flour
1 C. Whole Wheat flour
1 C. nuts

Mix together zucchini, oil, honey, eggs and then add dry ingredients and nuts. Pour into a loaf pan and bake at 325 degrees for 1 hr.

Pumpkin Bread

1 C. unbleached flour
2/3 C. Whole Wheat flour
¼ t. baking powder
1 t. soda
¾ t. salt
½ t. cinnamon
½ t. nutmeg
1/3 C. cooking oil
2/3 C. honey
½ t. vanilla
2 eggs
1/3 C. water
½ C. chopped nuts
1 C. mashed pumpkin

Grease loaf pan (9x5x3 inches). Sift flour and other dry ingredients. In a mixing bowl cream shortening, honey and vanilla. Add eggs one at a time beating thoroughly after each addition. Stir in pumpkin. Stir in dry ingredients alternately with water just until smooth. Fold in nuts and turn batter into prepared baking pan. Bake at 350 degrees for 45 to 55 min. or until done (test with cake tester inserted into center of loaf). It is done if tester comes out clean. Cool on rack.

Date and Nut Loaf

1 C. chopped dates
1 level t. soda
1 C. boiling water
1 tb. shortening
1 unbeaten egg
2/3 C. honey
1 C. chopped nuts
½ C. Whole Wheat flour
1 C. unbleached flour
½ t. salt

Put dates in a mixing bowl, sprinkle soda over them and 1 C. boiling water. Let stand until cool. Add 1 tb. shortening and 1 unbeaten egg and stir, add honey and nuts. Add flour and salt and stir. Put in a loaf tin and bake for 1 hr. at 325 degrees or until it can be pierced with a pick, which comes out clean.

Very Special Refrigerator Rolls

Dissolve 2 pkg. dry yeast in ¼ C. warm water. Put in mixing bowl. Add 1/3 C. honey, ½ C. cooking oil and 1 C. warm water. Add 4½ C. unbleached flour, ½ tb. salt and 3 beaten eggs. Let rise until double in bulk. Put in a greased bowl in refrigerator over night or for several hours. Take out of refrigerator 3 to 4 hours before serving time. Divide dough in half and roll out about ¼ in. thick on floured board and spread with butter. Roll like a jelly roll and cut in 1 in. slices. Place in greased muffin tins and let rise. Bake at 425 degrees until brown. This dough can be used for sweet rolls by adding cinnamon, nuts, and honey.

Company Rolls

1¼ C. evaporated milk
1 tb. salt
2 eggs
¼ C. honey
1 C. hot water
¼ C. warm water
1 dry yeast cake
6 to 8 C. sifted
unbleached flour

Sprinkle yeast over ¼ C. warm water and let stand to dissolve. Beat eggs, add honey, salt, milk and hot water, one at a time. Mix well. Add yeast mixture and about 5 C. flour gradually. Place on floured board and knead until smooth, adding more flour as needed. Roll ½ in. thick. Spread with butter, fold over and respread with butter. Do this 5 times. Then cut with cookie cutter and put in greased muffin tins. Let rise 2 to 3 hrs. Bake at 425 degrees for 10 to 12 min. Makes 32 rolls.

Whole Wheat Refrigerator Rolls

2 yeast cakes
½ C. warm water
1 C. milk
1/3 C. oil
¼ C. honey
2 eggs well beaten
1 t. salt
4½ to 5 C. Whole Wheat flour

Crumble yeast cakes into warm water and let sit. Scald milk and cool to luke warm then combine milk, oil, honey, salt and eggs. Add yeast mixture, then flour and mix well with a spoon. The mixture will be sticky. Put in refrigerator overnight. Two hours before rolls are to be served, take dough out of refrigerator and let stand at room temperature for about 30 min. Knead well on canvas-covered board. Roll out and cut for rolls. Place in greased pan and let rise about 1 hr. Bake 15 or 20 min. at 375 to 400 degrees.

Cheese Puffs (Muffins)

1 C. unbleached flour
1 C. grated cheddar cheese
1 t. baking powder
1 C. milk
1 egg
¼ tb. salt

Beat egg lightly and add flour and milk alternately. Mix in cheese and bake in greased muffin tins at 350 degrees for 15 to 20 min.

Oatmeal Muffins

Soak for 1 hr:
1 C. (quick cooking) rolled oats
1 C. buttermilk or sour milk

Sift together:
1 C. unbleached flour
1 t. baking powder
¼ t. soda

Mix 1/3 C. shortening
or oil (part butter)
1/3 C. honey
1 egg
½ t. salt

Stir in alternately with rolled oats and buttermilk. Fill greased muffin tins about 1/3 full. Bake at 400 degrees for 20 to 25 min.

Quick Bran Muffins

Put into a mixing bowl 1 C. bran, add 1 C. milk and let soak. Sift 1 C. unbleached flour, 2 t. baking powder and ½ t. salt. Add 1/3 C. cooking oil. Mix thoroughly and spoon into muffin tins about ½ full. Bake at 425 degrees for about 15 min. or until done.

Bran Muffins

1 C. bran
1 C. buttermilk or sour milk
1/3 C. cooking oil
1/3 C. honey
1 egg
1 C. unbleached flour
1 t. baking powder
½ t. soda
1 t. salt

Mix oil, honey and egg, and add dry ingredients alternately with buttermilk. Fill greased muffin tins 2/3 full and bake at 400 degrees for 20 to 25 min.

Biscuits

1 C. Whole Wheat flour
1 C. unbleached flour
2½ t. baking powder
1 t. salt
¼ C. shortening or
  cooking oil
¾ C. milk

Mix dry ingredients and cut in shortening. Add milk, then place dough on a floured board and cut into biscuits and bake in a greased pan at 450 degrees for 10 to 12 min.

Sprouted Wheat Pancakes

1 egg
1¼ C. buttermilk
½ t. soda
1¼ C. unbleached flour
1 t. honey
2 tb. cooking oil
1 t. baking powder
½ t. salt
½ C. sprouted wheat

Beat egg and add buttermilk and soda, then add flour, honey, cooking oil, baking powder and salt, add sprouted wheat. Bake on hot griddle.

Sour Dough Hotcakes

1 t. salt
1 tb. honey
1 t. soda
2 eggs
3 tb. melted shortening

In the evening or 6 or 8 hours before using, set the sponge batter as directed above. In the morning save ½ C. for next starter. Add salt, sugar, soda and eggs and blend well. Add melted shortening. Bake on hot griddle, turn once. For interesting variations add ½ C. Whole Wheat flour, cornmeal, wheat germ or bran flakes to batter. The 2 eggs in the recipe will provide the liquid for this addition.

Sour Dough Waffles

1 t. salt
1 tb. honey
1 t. soda
2 eggs
¼ C. melted shortening

Set sponge as directed, making it slightly thicker, and let stand overnight. Remove the usual starter for next time. To the remaining sponge add eggs and dry ingredients and mix well. Add melted shortening just before baking. Bake according to directions on waffle iron.

Sour Dough Muffins

1½ C. Whole Wheat flour
1/3 C. honey
1 t. salt
1 t. soda
1 C. raisins (optional)
½ C. cooking oil or shortening
1 or 2 eggs

Set sponge as usual, saving ½ C. for next starter. Sift ingredients into a bowl. Make a well in the center. Mix egg, shortening and honey thoroughly with sponge. Pour this into the well in the flour and stir only to moisten flour. Fill muffin tins ¾ full. Bake at 375 degrees for 30 or 35 min. Makes 12 large muffins.

Sour Dough Blueberry Muffins

½ C. Whole Wheat flour
1½ C. unbleached flour
¾ C. Sour Dough Starter
½ C. cooking oil
¾ t. baking soda
1/3 C. honey
½ C. undiluted canned milk
1 egg
1 C. drained blueberries
or other fruit

Mix dry ingredients, add oil and mix, add egg, milk, and blueberries. Add ¾ C. sour dough starter from the crock (or enough to make the mixture moist and hold together nicely). Do not beat vigorously. Bake in greased muffin tins for about 30-35 min. at 375 degrees. These muffins bake more slowly than usual so be sure they are done before removing from oven.

IX

Cakes and Cookies

Whole Wheat Spice Cake

½ C. Safflower or other oil
2/3 C. honey
2 eggs
¾ C. buttermilk
1 t. baking powder
1 t. soda
2½ C. Whole Wheat flour
1 t. cinnamon
¼ t. allspice
¼ tb. nutmeg
1 C. raisins, that have been
soaked in hot water and
thoroughly drained
½ t. salt
½ C. chopped nuts

Beat shortening, honey and eggs. Add buttermilk alternately with dry ingredients. Stir in raisins and nuts. Bake for about 30 min. at 350 degrees.

Banana Cake

2/3 honey
½ C. oil or shortening
2 C. Whole Wheat flour
1 t. baking soda
1 tb. buttermilk
½ t. salt
2 t. vanilla
1 C. mashed ripe bananas
1 C. nuts (optional)
2 eggs

Mix shortening and honey, add eggs and stir. Add bananas alternately with dry ingredients. Bake about 40 min. at 325 degrees.

Apple Cake

2/3 C. honey
½ C. oil or shortening
2 C. grated raw apple
2 C. Whole Wheat flour
1 t. salt
1 t. cinnamon
1 t. allspice
1½ t. soda
½ C. nuts

Cream shortening and honey, add apples alternately with dry ingredients, and add nuts. Bake for about 50 min. at 325 degrees or until a pick inserted comes out clean.

Rhubarb Cake

½ C. cooking oil
1 C. honey
1 beaten egg
1 C. buttermilk
1 t. soda
1½ C. unbleached flour
½ C. wheat germ
1½ C. cut up raw rhubarb
1 t. cinnamon

Cream together cooking oil and honey. Add beaten egg. Add alternately buttermilk, soda and flour. Fold rhubarb in lightly. Spread in greased and floured pan. Sprinkle top with cinnamon. Bake at 350 degrees.

Rhubarb Crumble

2½ C. cut fresh rhubarb
½ C. honey
1/3 C. unbleached flour
1 C. quick cooking oats
¾ C. margarine or butter
½ t. salt
3 tb. water
¾ C. honey
¾ C. unbleached flour

Mix together rhubarb, ½ C. honey, flour and salt. Place in buttered baking dish. Sprinkle with cinnamon. Mix separately ¾ C. honey, flour, oats, butter and salt. Pour over rhubarb mixture as topping. Sprinkle with water. Bake at 350 degrees for about 40 min. Serve warm or cold plain or with cream, ice cream or cheese.

Sour Dough Cake Carob (chocolate)

½ C. Starter
1 C. milk
1½ C. unbleached flour

Mix well and let stand 2 to 3 hours in a warm place until bubbly and there is a clean sour milk odor.

2/3 C. honey
½ C. shortening
¼ t. salt
1 t. vanilla
1 t. cinnamon
1½ t. soda
2 eggs
3 squares melted Carob (chocolate)

Cream shortening, honey, flavoring, salt, and soda. Add eggs one at a time, beating well after each addition. Combine creamed mixture and melted carob with sour dough mixture, stir 300 strokes or mix at low speed until blended. Pour into layer pans or one large pan. Bake at 350 degrees for 25 min. or until done. Cool and frost with carob frosting of your choice.

Applesauce Cake

½ C. butter
2/3 C. honey
1 egg
1 C. unsweetened applesauce
½ C. Whole Wheat flour
1½ C. unbleached flour
½ t. salt
¼ C. chopped nuts
½ t. baking powder
1 t. soda
½ t. cloves
1 t. cinnamon
1 t. allspice
1 C. chopped raisins

Grease well an 8 in. square cake pan. Dredge with flour and shake.

Cream butter, add honey, beat until light. Add the whole egg and applesauce. Beat thoroughly. Sift together flour and other dry ingredients and add to mixture. Add nuts and raisins (raisins may be soaked in hot water for a few minutes to make chopping less difficult, then drain off water). Pour into cake pan. Temperature 375 degrees; baking time 25 to 35 min.

Graham Cake

1 C. shortening or cooking oil
1 C. honey
1 sifter of Whole
Wheat flour
3 eggs
1 t. salt
2 t. soda
3 tb. carob powder
1 t. cinnamon
1 t. nutmeg
2 t. vanilla
1 box of raisins

Cream shortening and honey, add eggs. Put raisins into pan and cover with water; allow to come to a boil. Set aside to cool; drain. Add raisins and dry ingredients, which have been mixed together. Put in a large pan and bake at 350 degrees until done. Test by inserting a pick. When it comes out clean the cake is done.

Ginger Cookies

2½ C. unbleached flour
1 C. wholewheat flour
1 t. soda
¾ C. butter or cooking oil
1 t. salt
1 t. ginger
1 t. cinnamon
¼ C. honey
1 C. molasses
1/3 C. buttermilk

Mix dry ingredients. Put cooking oil, molasses and honey in mixing bowl and mix. Add dry ingredients alternately with buttermilk. Chill dough in refrigerator, then roll on floured board to ⅛ in. thickness. Cut with a cookie cutter and place on a greased baking sheet. Bake at 375 degrees for about 15 min. Makes 5 doz. cookies.

Gluten Cookies

1 C. cooking oil
2 eggs
1 t. vanilla
1 C. ground gluten
½ t. salt
2/3 C. honey
1 t. soda
1½ C. Whole Wheat flour
1 C. oatmeal
½ C. chopped nuts

Mix all ingredients and drop by spoonful on a greased cookie sheet. Bake at 350 degrees for about 10 min.

Molasses and Honey Cookies

¼ C. cooking oil or shortening
¼ C. butter
¼ C. honey
½ C. molasses (light)
1 egg slightly beaten
¼ C. milk
1 C. unbleached flour
1 C. Whole Wheat flour
½ t. salt
½ t. ginger
½ t. cloves
½ t. cinnamon
1 t. baking soda

Cream the first six ingredients together. Add the dry ingredients to the first mixture and drop by teaspoonsful on a greased cookie sheet and bake 10 min. at 375 degrees.

Rolled Molasses and Honey Cookies

Add about ½ C. flour to the above recipe to make rolled cookies.

Honey Peanut Butter Cookies

Mix together:
- ¼ C. butter
- ½ C. peanut butter (that contains just peanuts)
- 2/3 C. honey
- 1 egg

Sift together 1 C. unbleached flour
- ½ C. Whole Wheat flour
- ½ t. baking powder
- ¾ t. soda
- ¼ t. salt

Stir the dry ingredients into the creamed mixture. Chill dough. Roll into balls the size of walnuts and place about 3 inches apart on a lightly greased cookie sheet. Flatten with a fork that has been dipped in flour—criss-cross. Bake until set but not hard—375 degrees for about 10 min.

Carrot Cookies

¾ C. Safflower or other
natural oil
2/3 C. honey
1 egg
1 t. baking powder
1 t. vanilla
1½ C. finely shredded
raw carrots
2 C. Whole Wheat flour
½ t. salt
1 t. cinnamon
½ t. cloves
½ t. nutmeg
1 C. chopped walnuts

Mix honey, shortening and egg together. Add dry ingredients, carrots and vanilla and mix. Add nuts. Drop from a teaspoon on a greased cookie sheet and bake for about 15 min at 375 degrees. (You may use shortening if you desire.)

Pineapple, Coconut Cookies

2 C. Whole Wheat flour
2 t. baking powder
½ t. salt
2/3 C. Safflower or other
natural oil
¾ C. honey
½ t. almond flavoring
½ t. vanilla flavoring
1 egg
½ C. drained crushed pineapple
½ C. shredded coconut, fine

Mix oil, honey and egg. Beat well. Add dry ingredients alternately with pineapple and coconut. Drop on greased cookie sheet and bake at 325 degrees for about 15 or 20 min. (You may use shortening stead of oil.)

Prune Cookies

2/3 C. honey
1 C. butter or oil
2½ C. Whole Wheat flour
1 egg
1 C. chopped cooked prunes
2 t. baking powder
1 t. vanilla
½ t. cloves
½ t. cinnamon
1 t. soda

Mix honey, shortening, egg and prunes. Add dry ingredients and vanilla. Drop from teaspoon and bake for about 15 min. at 350 degrees.

X
Casseroles and Soups

Ground Beef and Vegetable Casserole

2 lbs. of lean ground beef
½ t. sage
2 t. salt
10 onions fried
Buttered crumbs
1 C. chopped celery
1 C. cooked or sprouted wheat
4 tb. cooking oil
3 tb. unbleached flour
2 C. unstrained tomatoes

Make medium-sized balls of the ground beef and place in the bottom of a casserole. Season with sage and salt. Over this cover with wheat, celery and fried onion. Mix the oil and flour, add tomatoes and cook until thickened. Pour this over the onions. Cover with buttered crumbs and bake uncovered at 375 degrees for 1½ hr.

Chicken and Wheat Casserole

1 Can (10½ oz.) cream
of mushroom soup
1 C. chopped celery, cook
until tender
½ C. mayonnaise or
salad dressing
3 tb. chopped pimento
2 tb. chopped onion
cooked until tender
½ C. or more sprouted or
cooked wheat
2 C. cooked turkey
or chicken
½ C. chopped blanched
almonds
1 tb. lemon juice
3 chopped hard
cooked eggs (optional)
½ t. salt

Combine in baking dish and sprinkle with topping and bake at 375 degrees for 30 min. or until bubbling hot. Makes 6 servings.

Topping

Combine ½ C. unbleached flour, ½ C. sesame seeds, ½ C. shredded mild cheddar cheese, ¼ t. salt, and ¼ C. soft butter.

Meat Loaf

1 lb. ground beef (lean)
½ C. chopped onion
½ C. chopped celery
½ t. salt
¼ t. garlic salt
½ C. cooked or sprouted
wheat, rolled wheat, or
whole wheat bread crumbs
1 egg
about ¼ C. water

Mix and place in a loaf tin that has been greased and bake at 325 degrees for 1 hr. or until it leaves the side of the pan and is brown on top.

This mixture can be used for ground beef patties or meat balls. If you desire, a can of cream of mushroom soup that has been diluted with ½ C. water and ½ C. condensed milk may be poured over the meat balls or patties and simmered for about half an hour. Use salad dressing to brown meat balls or patties instead of shortening for a pleasant change.

Chicken Gluten Pie

2 C. chicken gluten
(see basic recipe)
1 C. chopped cooked carrots
1 C. chopped cooked celery
1½ qts. chicken broth
½ C. chopped cooked onion
1 C. diced cooked potatoes
1 C. cooked peas

Add all ingredients and simmer ½ hr. Thicken with 3 level tb. flour that has been mixed with chicken broth. ¼ C. chopped parsley may be added. Cover with a rich biscuit dough and bake at 400 degrees until brown.

Biscuit Dough (Rich)

2 C. unbleached flour
2½ t. baking powder
1 t. salt
1/3 C. cooking oil
¾ C. milk

Mix all ingredients, stirring in milk last. Put on a floured board and roll to desired thickness.

Gluten Meat Loaf

2 C. ground gluten
2 tb. Whole Wheat flour
2 tb. cooking oil
1 medium-sized
    onion chopped
3 C. water, to which
    has been added 2
    bouillon cubes
½ C. chopped celery
2 eggs
1 t. seasoned salt
¼ t. garlic salt

Add 2 bouillon cubes to 3 C. water and boil until cubes are dissolved. Add ground gluten and simmer ½ hr. and then let soak 1 to 1½ hr. before using; drain. Then add and mix all other ingredients. Place in a greased loaf tin and bake 30 to 40 min. at 325 degrees.

Gluten Beef Stew

Cut baked gluten into cubes, 1 C. for a small stew and 2 C. for a larger one. Place 3 bouillon cubes in a quart of water, add the cubed gluten and let simmer for about 30 min. and let stand for about 1½ to 2 hrs. Drain. In a separate pan put the bouillon water and bring to a boil. Add about ½ C. chopped onion, 1 C. chopped celery. Cut up potatoes, chopped carrots and any other vegetable you desire and cook until tender. Add the cubed gluten. Season with seasoned salt. Add water if needed and thicken by adding 2 tb. flour that have been mixed with ½ C. water. Add salt if needed, and serve.

Basic Chicken Gluten

1 qt. of water
3 chicken bouillon cubes
1/3 t. poultry seasoning

Cube 2 C. baked gluten and place in pan and simmer for 30 min. Soak for 1 to 2 hrs.

This is ready to be used in a variety of ways.

Chicken Loaf

3 C. chicken gluten
that have been ground or cut
in small pieces
1½ C. soft Whole Wheat
bread crumbs
4 oz. sliced mushroom pieces
½ t. salt
¼ C. finely cut pimento
1 C. milk
1 C. chicken broth
2 eggs
¼ C. chopped parsley
¼ C. chopped onion

Put this mixture in a greased baking dish and bake 50 min. at 350 degrees. Slice and serve hot.

Ham Gluten

Gluten may be made to taste like ham by using baked gluten that has been ground, sliced or cubed; simmer in water to which has been added ham flavoring. Soak 1½ hr. Drain and use as desired.

Ground Beef and Sprouted Wheat Casserole

1 C. chopped carrots
1 C. chopped celery
2 C. cooked green beans
½ to 1 C. sprouted
or cooked wheat
1 lb. lean ground beef
1 medium-sized onion
chopped fine
1 beef cube
2 tb. flour
6 medium potatoes
boiled and mashed

Brown the ground beef and onion in a frying pan. Cook celery and carrots in the water you use to cook the green beans or the liquid in a can of beans. Make a gravy by dissolving the beef cube in 1½ C. hot water and thicken with thickening made with 2 tb. flour. Mix the gravy, vegetables and ground beef together and season to taste with salt, garlic salt and "Season All" salt. Cover with mashed potatoes and bake in the oven until hot about 25 min.

Chicken Dressing Casserole

6 C. soft Whole Wheat bread crumbs
½ C. butter
1 t. salt
1 C. chopped celery cooked
1 chopped medium onion cooked
1 t. ground sage

Use a 3 to 4 lb. stewing hen or frying chicken. Cover with water, add some cut up celery and onion and 1 t. salt and cook until tender. Remove the chicken from the broth and cool. Cut up in bite-size pieces.

Mix the bread crumbs and other ingredients and place in the bottom of a greased baking dish. Place the chicken on top and pour the sauce over the top of it. Bake at 300 degrees for 1 hr., covered for part of the baking.

Sauce

½ C. chicken fat
1 C. unbleached flour
1 C. milk
4 C. chicken broth
2 t. salt
4 eggs

Melt fat in pan and blend in flour. Add milk gradually, stirring constantly. Cook until thickened. Beat and pour thickened mixture over eggs, stirring until blended. Return to pan and heat but do not boil. Pour over casserole. Top with buttered whole wheat bread crumbs and bake as directed.

Vegetable Casserole

6 to 8 tomatoes (fresh
or canned)
1 head of cabbage,
shredded
3 zucchini sliced
1 lb. broccoli, cut
in pieces
1 bunch of green onions,
sliced
5 carrots, sliced
2 C. chopped celery
1 C. cooked or sprouted
wheat
2 C. grated mild
cheddar cheese

Put tomatoes in blender. Layer the vegetables with cheese in a large casserole. Pour tomatoes over vegetables to barely cover. Cover and bake at 350 degrees for 30 to 45 min. or until vegetables are done as you desire. Makes 8 servings. Other vegetables may be used.

Vegetable Loaf

12 oz. of protein
powder (soy)
water
2 eggs
½ C. diced celery
½ C. diced carrots
½ C. chopped onion
½ C. cooked or
sprouted wheat
1 clove garlic, minced

Mix soy powder with water as directed on package, mix together with eggs and all other ingredients and pack into a well greased loaf pan. Bake as directed on soy package. Makes 6 servings.

Salmon and Wheat Patties

1 lb. can of salmon
1 small onion
½ C. sprouted or
cooked wheat
1 egg
soda crackers

Flake salmon, add onion and egg and mix, add cooked or sprouted wheat, liquid from salmon, and enough soda cracker crumbs or whole wheat bread crumbs to make patties that are easily handled. Brown in a frying pan with 1 tb. cooking oil or bake in a pan in oven at 325 degrees for 30 to 40 min.

Chicken Chow Mein Casserole

One 3 lb. chicken
1½ C. chopped celery
2/3 C. chopped onion
2/3 C. cooked or
    sprouted wheat
1 Can cream of mushroom
    soup
1½ C. milk
1 large can of
    Chow Mein noodles

Cook chicken and remove bones, cut into bite-sized pieces. Saute onions and celery in butter. Put Chow Mein noodles in a casserole dish. Mix the above ingredients and pour over noodles. Bake at 350 degrees for 30 or 40 min.

Tuna Broccoli Casserole

5 eggs
½ C. milk
Two 6½ oz. cans of
tuna, drained
1 tb. lemon juice
1 C. Whole Wheat bread
crumbs
½ t. salt
½ C. grated cheddar
cheese
1½ C. broccoli cooked
and drained
½ C. cooked or sprouted
wheat

Beat eggs lightly, add milk and bread crumbs. Let stand for a few minutes. Stir in lemon juice, onion, salt, tuna and cheese. Put cooked broccoli and wheat in blender, cover and beat until smooth. Mix this into the other ingredients. Place in a loaf tin that has been greased and bake at 350 degrees for 1 hr. Turn out on a serving dish. Garnish with parsley.

Golden Burgers

One 3½ oz. French
Fried onions
2 lb. lean ground beef
1 egg beaten
1 tb. Worcestershire
sauce
1 t. salt
½ t. garlic salt
6 large hamburger buns or
6 large oval Italian
bread slices, ¾ in. thick
1 C. cooked, cooled bulgar or
1½ C. Whole Wheat
bread crumbs
Cheese sauce
6 tomato slices
green pepper rings (optional)

Mix ground beef, onions, egg, seasoning, bulgar of crumbs. Broil, cook in oven or in a frying pan until done to your taste. Place burger on toasted bun or bread and cover with cheese sauce. Garnish with onion rings, tomato and green pepper.

Cheese Sauce

2 tb. butter or margarine
2 tb. flour
½ t. salt
1 C. milk
1½ C. grated cheddar cheese
½ t. prepared mustard
1 t. Worcestershire sauce

Melt butter in heavy sauce pan, stir in flour and seasoning. Add milk and cook. Stir until thickened and cheese melts. Makes about 1½ C.

Ground Beef and Vegetable Soup

1½ lbs. of lean ground beef
1 C. chopped onions
1 C. chopped celery
1 qt. water
1 C. chopped carrots
1 C. chopped cabbage
1 C. chopped parsley
2 t. sweet basil
½ t. celery salt
½ t. garlic salt
1 C. chopped green pepper
1 C. chopped zucchini squash
1 C. cooked whole wheat
1 C. chopped cauliflower
1 C. chopped mushrooms
1 C. cut green beans
2 qt. tomatoes that have
been chopped
½ t. onion salt

In a large kettle, lightly brown ground beef. Stir to break into small pieces. Add celery and onion, stir and add water. Simmer for 5 min. Add the other vegetables and seasoning. The parsley and mushrooms may be added last. All vegetables should be fresh if possible. Canned or frozen vegetables can be used if you cannot obtain the fresh ones. Other vegetables that may be chopped and used are potatoes, turnips, peas and corn. Add more seasoning if you desire. This makes a large pan of soup. Part of it may be frozen to be used at another time if you wish. This is a thick soup.

Chicken Vegetable Soup

Cook a 3 lb. chicken in water with ½ tb. salt until tender. Remove the skin and bones and cut in bite-size pieces. Set aside.

In chicken broth (about 3 qts.) cook:

1 C. chopped onions
1 C. chopped carrots
2 C. chopped green pepper
1 C. cooked whole wheat
1 C. chopped mushrooms
2 C. chopped celery
1 C. chopped cabbage
1 C. chopped zucchini squash
1 C. cauliflower
1 C. chopped parsley

Add the vegetables to the chicken broth and cook until tender, add chicken and simmer about 10 min. The vegetables should be fresh. Other vegetables may be added if you desire.

XI
Salads

Chicken Gluten Salad

2 C. cubed chicken gluten
prepared as in basic recipe
2 tb. finely chopped onion
2 C. chopped celery
½ C. blanched, chopped
almonds
½ C. grated mild cheddar
cheese

Moisten with salad dressing. The salad may be prepared several hours before it is to be served.

Chicken Gluten Spring Salad

2 C. cubed chicken gluten
prepared as in basic recipe
2 tb. finely chopped onion
1 C. chopped celery
1 medium sliced tomato
1 C. fresh or frozen green
peas that have been cooked
and cooled (don't overcook)
1 sliced cucumber
1 medium head of lettuce
that has been washed, dried
and broken in pieces

Use any other vegetable that you desire. Mix and serve with salad dressing.

Chicken Gluten Fruit Salad

2 C. cubed chicken gluten
as prepared in basic recipe
1 C. chopped celery
2 C. diced red delicious
apples (leave skin on)
1 C. pineapple tid bits
(canned in own juice)
1 C. sliced bananas
Juice of 1 lemon
1 C. unsweetened pineapple
juice
1 egg
1 tb. corn starch

Stir some of the juice in the corn starch then add beaten egg and the rest of the juice. Put in a pan and stir over burner until thick. Cool. Mix in the fruit and chicken gluten. Chill and serve on a lettuce leaf. (If this is not sweet enough to suit you, add some honey to the thickened juice mixture.)

Shrimp Salad

1½ lb. loaf of 50%
whole wheat bread
4 hard boiled eggs
2 lbs. of shrimp
2 C. celery finely chopped
2½ C. mayonnaise
1/3 C. finely chopped
green onions

Slice bread and cut off crusts; cube. Mix onion and cut up eggs with bread and let stand in the refrigerator over night. Next day mix all ingredients and serve on a lettuce leaf.

Tuna Fish and Wheat Salad

1 C. cooked or sprouted wheat
1 C. chopped celery
½ C. chopped green pepper
¼ C. chopped green onion
One 6½ or 7 oz. can
of tuna fish
Albacore preferred (drain
off liquid and flake)
½ C. shredded carrot

Mix together with salad dressing to taste and serve on a lettuce leaf.

Cottage Cheese and Sprouted Wheat Salad

Mix equal parts of cottage cheese, crushed pineapple and sprouted wheat. Serve on a lettuce leaf with salad dressing.

Tossed Salad

Add sprouted wheat to any salad. For a vegetable salad use crisp lettuce, that has been broken into pieces and any of the following: tomato wedges, a little chopped onion, sliced cucmbers, red apples unpeeled and diced, diced zucchini, orange segments, green peas (raw or slightly cooked), cauliflower, celery, radishes, cooked diced beets that have been soaked in a little vinegar water. Use salad dressing of your choice.

Cabbage Salad

2 C. chopped crisp cabbage that has been washed and drained. Add 1 large grated carrot and ½ C. sprouted wheat. The amounts depend on the size of a salad that you wish. Use salad dressing or mayonnaise. If you wish, add raisins that have been soaked in warm water. Almond, cashew or peanuts are good added to this salad. Red diced apples may be used in place of carrots.

XII
Sprouted Wheat Recipes

Sprouted Wheat and Tuna Sandwich

Drain the oil from a can of tuna fish and stir with a fork. Add salad dressing and ½ C. sprouted or cooked wheat. Spread the tuna mixture over a slice of bread, place a slice of mild cheddar cheese on top and cover with another slice of bread. Butter the outside of the bread on both sides and brown in a medium hot frying pan. Serve hot.

Sprouted Wheat Cereal

1 C. sprouted wheat
2 C. boiling water
½ t. salt

Cook over low heat for 30 min., then let stand for 10 min. Serve with cream and honey as a breakfast cereal. It is also good served with butter or gravy. To use as cereal chopped dates or raisins may be added.

Quick Scrambled Eggs with Sprouted Wheat

Pour eggs into moderately hot fry pan to which butter has been added. Put in sprouted wheat, about 1 tb. per egg. Add a pinch of salt and stir gently until thick.

Frozen Mixed Vegetables and Sprouted Wheat

Add ½ C. or more of sprouted wheat to 1 package of frozen vegetables, 1 tb. butter and cook as directed on the frozen vegetable package.

XIII
Miscellaneous

Parched or Popped Wheat

Soak wheat for 1½ hrs. Drain and dry by putting in a warm oven and stirring occasionally. Put in a heavy fry pan with a little cooking oil in it. Stir or shake pan until wheat has popped. Remove and season with salt and butter. (It won't pop into large flakes like pop corn.)

Dressing for Turkey or Other Meats

2 C. Whole Wheat bread crumbs (more if desired)
1 C. chopped cooked celery
½ C. chopped onion
½ t. ground sage

Use enough chicken broth to moisten. Put this in the turkey or by the side of it when it has about 1 hr. of cooking time left, or cook in a separate pan in the oven. The pan should be greased so the dressing won't stick.

Pie Crust

1 C. unbleached flour
1 C. Whole Wheat flour
2/3 C. cooking oil or
shortening
1 t. salt

Mix the above ingredients with a spoon or two knives. Moisten with a little cold water until the dough will stick together enough to roll.

Bran Flakes

2 C. thin raw bran
(with water)
½ t. salt
2 tb. honey

Mix and pour on an oiled cookie sheet and bake at 300 degrees for about 20 or 25 min. Take from pan immediately. Cool and break in pieces.

Steam Pudding

2 C. Whole Wheat
bread crumbs
1 C. milk
2/3 C. honey
2 tb. unbleached flour
1 t. soda
1 t. cinnamon
1 t. nutmeg
½ t. cloves
1/3 C. cooking oil
1 C. chopped dates
1 C. nuts

Mix all ingredients together and steam for 2 hrs. Serve with light cream or nutmeg sauce.

Nutmeg Sauce

1/3 C. unbleached flour
1/3 C. honey
¼ lb. butter
2 C. water
½ t. nutmeg

Mix flour, butter, nutmeg and honey. Gradually add water and bring to a good boil.

INDEX

Miscellaneous

Muffins, Rolls, etc.

Salads

Sourdough